Table of Contents

PREFACE

On March 28, 2014 the Obama Administration released a key element called for in the President's Climate Action Plan: a Strategy to Reduce Methane Emissions. The strategy summarizes the sources of methane emissions, commits to new steps to cut emissions of this potent greenhouse gas, and outlines the Administration's efforts to improve the measurement of these emissions. The strategy builds on progress to date and takes steps to further cut methane emissions from several sectors, including the oil and natural gas sector.

This technical white paper is one of those steps. The paper, along with four others, focuses on potentially significant sources of methane and volatile organic compounds (VOCs) in the oil and gas sector, covering emissions and mitigation techniques for both pollutants. The Agency is seeking input from independent experts, along with data and technical information from the public. The EPA will use these technical documents to solidify our understanding of these potentially significant sources, which will allow us to fully evaluate the range of options for cost-effectively cutting VOC and methane waste and emissions.

1.0 INTRODUCTION

The oil and natural gas exploration and production industry in the U.S. is highly dynamic and growing rapidly. Consequently, the number of wells in service and the potential for greater emissions from oil and natural gas sources is also growing. There were an estimated 504,000 producing gas wells in the U.S. in 2011 (U.S. EIA, 2012a), and an estimated 536,000 producing oil wells in the U.S. in 2011 (U.S. EIA, 2012b). It is anticipated that the number of gas and oil wells will continue to increase substantially in the future because of the continued and expanding use of horizontal drilling combined with hydraulic fracturing (referred to here as simply hydraulic fracturing).

Due to the growth of this sector and the potential for increased air emissions, it is important that the U.S. Environmental Protection Agency (EPA) obtain a clear and accurate understanding of emerging data on emissions and available mitigation techniques. This paper presents the Agency's understanding of emissions and available emissions mitigation techniques from a potentially significant source of emissions in the oil and natural gas sector.

1.1 Definition of the Source

The focus of this white paper is natural gas-driven pneumatic controllers and natural gas-driven pneumatic pumps. Such pneumatic controllers and pumps are widespread in the oil and natural gas industry and emit natural gas, which contains methane and VOCs. In some applications, pneumatic controllers and pumps used in this industry may be driven by gases other than natural gas and, therefore, do not emit methane or VOCs.

1.1.1 Pneumatic Controllers

For the purposes of this white paper, a *pneumatic controller* means an automated instrument used for maintaining a process condition such as liquid level, pressure, pressure difference and temperature. Based on the source of power, two types of pneumatic controllers are defined for this paper:

2

- *Natural gas-driven pneumatic controller* means a pneumatic controller powered by pressurized natural gas.
- *Non-natural gas-driven pneumatic controller* means an instrument that is actuated using other sources of power than pressurized natural gas; examples include solar, electric, and instrument air.

Natural gas-driven pneumatic controllers come in a variety of designs for a variety of uses. For the purposes of this white paper, they are characterized primarily by their emissions characteristics:

- *Continuous bleed pneumatic controllers* are those with a continuous flow of pneumatic supply natural gas to the process control device (e.g., level control, temperature control, pressure control) where the supply gas pressure is modulated by the process condition, and then flows to the valve controller where the signal is compared with the process set-point to adjust gas pressure in the valve actuator. For the purposes of this paper, continuous bleed controllers are further subdivided into two types based on their bleed rate:
  - *Low bleed*, having a bleed rate of less than or equal to 6 standard cubic feet per hour (scfh).
  - *High bleed*, having a bleed rate of greater than 6 scfh.
- *Intermittent pneumatic controller* means a pneumatic controller that vents non-continuously. These natural gas-driven pneumatic controllers do not have a continuous bleed, but are actuated using pressurized natural gas.
- *Zero bleed pneumatic controller* means a pneumatic controller that does not bleed natural gas to the atmosphere. These natural gas-driven pneumatic controllers are self-contained devices that release gas to a downstream pipeline instead of to the atmosphere.

1.1.2   Pneumatic Pumps

Pneumatic pumps are devices that use gas pressure to drive a fluid by raising or reducing the pressure of the fluid by means of a positive displacement, a piston or set of rotating impellers. Pneumatic pumps are generally used at oil and natural gas production sites where electricity is not readily available (GRI/EPA, 1996d). The supply gas for these pumps can be compressed air, but most often these pumps use natural gas from the production stream (GRI/EPA, 1996e).

1.2   Background

1.2.1   Pneumatic Controllers

Pneumatic controllers are automated instruments used for maintaining a process condition such as liquid level, pressure, pressure differential, and temperature. In many situations, across all segments of the oil and gas industry, pneumatic controllers make use of the available high-pressure natural gas to operate control of a valve. In these natural gas-driven pneumatic controllers, natural gas is released with every actuation of the valve, i.e., valve movement. In some designs, natural gas is also released continuously from the valve control pilot. The rate at which the continuous release occurs is referred to as the bleed rate. Bleed rates are dependent on the design and operating characteristics of the device. Similar designs will have similar steady-state rates when operated under similar conditions. There are three basic designs of natural gas-driven pneumatic controllers: (1) continuous bleed controllers are used to modulate flow, liquid level, or pressure, and gas is vented continuously at a rate that may vary over time; (2) intermittent controllers release gas only when they open or close a valve or as they throttle the gas flow; and (3) zero bleed controllers, which are self-contained devices that release gas to a downstream pipeline instead of to the atmosphere (EPA, 2011a).

As noted above, intermittent controllers are devices that only emit gas during actuation and do not have a continuous bleed rate. Thus, the actual amount of emissions from an intermittent controller is dependent on the amount of natural gas vented per actuation and how often it is actuated. Continuous bleed controllers also vent an additional volume of gas during actuation, in addition to the device's continuous bleed stream. Thus, actual emissions from a continuous bleed device also depend, in part, on the frequency of activation and the amount of gas vented during activation. As the name implies, zero bleed controllers are considered to emit no natural gas to the atmosphere (EPA, 2011a).

In general, intermittent controllers serve functionally different purposes than bleed controllers and, therefore, cannot replace bleed controllers in most (but not all) applications. Furthermore, zero bleed controllers are "closed loop" systems that can be used only in applications with very low pressure and therefore may not be suitable to replace continuous bleed pneumatic controllers in many applications.

Non-natural gas-driven pneumatic controllers can be used in some applications. These controllers can be mechanically operated or use sources of power other than pressurized natural gas, such as compressed "instrument air." Instrument air systems are feasible only at oil and natural gas locations that have electrical service sufficient to power an air compressor. At sites without electrical service sufficient to power an instrument air compressor, mechanical or electrically powered pneumatic controllers can be used. Non-natural gas-driven controllers do not directly release methane or VOCs, but may have secondary impacts related to generation of required electrical power (EPA, 2011a).

1.2.2   Pneumatic Pumps

There are two types of pneumatic pumps that are commonly used in the oil and natural gas sector: piston and diaphragm (GRI/EPA, 1996d). These pumps have two major components, a driver side and a motive side, which operate in the same manner but with different reciprocating mechanisms. Pressurized gas provides energy to the driver side of the pump, which operates a piston or flexible diaphragm to draw fluid into the pump. The motive side of the pump delivers the energy to the fluid being moved in order to discharge the fluid from the pump. The natural gas leaving the exhaust port of the pump is either directly discharged into the atmosphere or is recovered and used as a fuel gas or stripping gas (GRI/EPA, 1996d).

The majority of pneumatic pumps used in oil and natural gas production are used for chemical injection or glycol circulation (GRI/EPA, 1996d). Pneumatic pumps used for chemical injection are needed in oil and natural gas production to inject small amounts of chemicals to limit processing problems and protect equipment. Typical chemicals that are injected into the process include: biocides, demulsifiers, clarifiers, corrosion inhibitors, scale inhibitors, hydrate inhibitors, paraffin dewaxers, surfactants, oxygen scavengers and hydrogen sulfide scavengers (GRI/EPA, 1996d). These chemicals are normally injected using pneumatic pumps at the wellhead, and into gathering lines or at production separation facilities (GRI/EPA, 1996d). Pneumatic pumps, commonly referred to as "Kimray" pumps, used for glycol circulation recover energy from the high-pressure rich glycol/gas mixture leaving the absorber and use that energy to pump the low-pressure lean glycol back into the absorber (GRI/EPA, 1996e).

1.3   Purpose of the White Paper

This white paper provides a summary of the EPA's understanding of the emissions from natural gas-driven pneumatic controllers and pumps in the oil and natural gas sector, the mitigation techniques available to reduce these emissions, the efficacy of these techniques and the prevalence of these techniques in the field. Section 2 of this document provides the EPA's understanding of emissions from pneumatic controllers and pumps, and Section 3 provides our understanding of available mitigation techniques. Section 4 summarizes the EPA's understanding based on the information presented in Sections 2 and 3, and Section 5 presents a list of charge questions for reviewers to assist the EPA with obtaining a more comprehensive understanding of pneumatic controller and pump VOC and methane emissions and emission mitigation techniques.

2.0   AVAILABLE EMISSIONS DATA AND ESTIMATES

There are a number of studies that have been published that have estimated VOC and methane emissions from pneumatic controllers and pneumatic pumps in the oil and natural gas sector. These studies have used different methodologies to estimate these emissions including the use of equipment counts and emission factors and direct measurement of emissions. Section 2.1 discusses the studies relevant to pneumatic controllers, and Section 2.2 discusses the studies relevant to pneumatic pumps. These studies are listed in Table 2-1, along with an indication of the type of information contained in the study.

Table 2-1. Summary of Major Sources of Pneumatic Controller and Pump Information

Report Name	Affiliation	Year of Report	Activity Factor	Pneumatic Controllers	Pneumatic Pumps
Methane Emissions from the Natural Gas Industry (GRI/EPA, 1996c)	Gas Research Institute / EPA	1996	Nationwide	X	X
Estimates of Methane Emissions from the U.S. Oil Industry (ICF Consulting, 1999)	EPA	1999	Nationwide	X	
Inventory of Greenhouse Gas Emissions and Sinks: 1990-2012 (U.S. EPA, 2014)	EPA	2014	Nationwide/ Regional	X	X
Greenhouse Gas Reporting Program (U.S. EPA, 2013)	EPA	2013	Basin	X	X
Measurements of Methane Emissions from Natural Gas Production Sites in the United States (Allen et al., 2013)	Multiple Affiliations, Academic and Private	2013	Nationwide	X	
Determining Bleed Rates for Pneumatic Devices in British Columbia (Prasino, 2013)	The Prasino Group	2013	British Columbia	X	
Air Pollutant Emissions from the Development, Production, and Processing of Marcellus Shale Natural Gas (Roy et al., 2014)	Carnegie Mellon University	2014	Regional (Marcellus Shale)	X	
Economic Analysis of Methane Emission Reduction Opportunities in the U.S. Onshore Oil and Natural Gas Industries (ICF, 2014)	ICF International	2014	Nationwide	X	X

2.1   Discussion of Data Sources for Pneumatic Controllers

This section presents and discusses pertinent studies and data sources that estimate emissions from pneumatic controllers.

2.1.1   Methane Emissions from the Natural Gas Industry (GRI/EPA, 1996c)

This report's main objective was to quantify annual methane emissions from pneumatic controllers from the natural gas production, processing, transmission, and distribution sectors. The methane emissions were determined by developing average annual emissions factors for the various types of pneumatic controllers used in each of the natural gas segments. The annual emission factors were then extrapolated to a national estimate using activity factors for each of the natural gas segments.

Production

The data used to develop emission factors for pneumatic controllers in the natural gas production sector were obtained from a study performed by the Canadian Petroleum Association (CPA)[1], manufacturers' data, measured emission rates[2], data collected from site visits, and literature data for methane composition.

The CPA study consisted of methane and VOC emission measurements from pneumatic controllers in two types of service: 19 in on/off service and 16 in throttling service.[3] The CPA study determined the average natural gas emission rate for on/off controllers was 213 standard cubic feet per day per device (scfd/device), and the average natural gas emission rate for throttling controllers was 94 scfd/device. For throttling controllers, the CPA study did not distinguish between the throttling controllers with intermittent bleed rates and throttling controllers with continuous bleed rates. In addition, only one throttling controller actuated during the emission measurement. Therefore, these measurements are lower in comparison to field measurements of similar devices in the U.S. (GRI/EPA, 1996c).

---

[1] Picard, D.J., B.D. Ross, and D.W.H. Koon, *A Detailed Inventory of CH$_4$ and VOC Emissions from Upstream Oil and Gas Operations in Alberta*. Canadian Petroleum Association, Calgary, Alberta, 1992.

[2] Controller survey data provided by Tenneco Gas Transportation, 1994 and Chevron, 1995.

[3] Controllers in on/off service wait until a specific set point is reached before actuating (e.g., a high or low liquid level). Controllers in throttling service maintain a desired set point (e.g., pressure).

The manufacturers' data were obtained from four manufacturers of pneumatic controllers and were based on laboratory testing of new controllers. The manufacturers' noted that emissions in the field can be higher due to operating condition, age, and wear of the device. The gas consumption rates for the manufacturers' pneumatic controllers ranged from 0 to 2,150 scfd. The manufacturers noted that the emissions from these controllers in the field may be higher than the reported maximum value (GRI/EPA, 1996c).

The measured emissions data[4] were collected by connecting a flow meter to the supply line between the pressure regulator and the controller to measure the gas consumption of the controller. The duration of the test depended on the operating conditions. For steady operating conditions, one data point was measured for 15-20 minutes. For variable operating conditions, several one-hour measurements were taken. The data set contained a total of 41 measurements from a combination of continuous bleed controllers from offshore and onshore production sites and transmission stations. The average gas emissions rates for continuous bleed controllers were determined to be 872 scfd/device for onshore and offshore production sites and 1,363 scfd/device for transmission stations.

The measured emission data[5] also provided data for intermittent bleed controllers that were measured using the same techniques that were used for the continuous bleed pneumatic controllers. A total of seven measurements were performed on intermittent bleed controllers located at onshore natural gas production sites. No measurements were available for intermittent bleed controllers in the offshore or transmission segments. The average natural gas emission rate for the intermittent pneumatic controllers was determined to be 511 scfd/device.

Site visit data were collected from a total of 22 sites to determine the number of pneumatic controllers located at natural gas production sites, and to determine the fraction of these controllers that were intermittent or continuous bleed. The study determined that 65% of

---

[4] Controller survey data provided by Tenneco Gas Transportation, 1994 and Chevron, 1995.
[5] Controller survey data provided by Tenneco Gas Transportation, 1994 and Chevron, 1995.

the pneumatic controllers were intermittent bleed and 35% of the pneumatic controllers were continuous bleed.

The measured emission data, the CPA study emissions data, and the pneumatic controller counts were used to develop a single emission factor for a "generic" pneumatic device. For the production segment, the "generic" pneumatic controller emission factor was calculated using:

- 323 scfd/device for intermittent bleed controllers,
- 654 scfd/device for continuous bleed controllers,
- a methane content of 78.8%, and
- the ratio of intermittent bleed to continuous bleed controllers at natural gas production sites.

The "generic" emission factor was determined to be 345 scfd/device of methane for a pneumatic controller at natural gas production sites.

Transmission

The transmission "generic" emission factor was calculated using data from three types of gas-operated pneumatic controllers: continuous bleed controllers and two types of intermittent bleed controllers used to operate isolation valves[6] (isolation valves with turbine operators and isolation valves with displacement-type pneumatic/hydraulic operators). The continuous bleed emission factor was obtained from the transmission station measured emission data, which was determined to be 1,363 scfd/device. The isolation valve with displacement-type pneumatic/hydraulic operators emission factor was determined using data provided by Shafer Valve Operating Systems[7,8] and the count of the isolation valves at four sites. Using these data,

---

[6] Isolation valves at transmission stations are very large and are most often actuated either pneumatically or by electric motor. Isolation valve pneumatic controllers only discharge gas when they are actuated and are considered to be intermittent.

[7] Shafer Valve Operating Systems. Gas Consumption Calculation Method for Rotary Vane, Gas/Hydraulic Actuators. Technical Bulletin Data, Bulletin GC-00693. June 1993.

the average annual emission factor was determined to be 5,627 standard cubic feet per year per device (scfy/device). For turbine-operated isolation valves, the natural gas emissions were estimated using information provided by Limitorque Corporation[9] and information from two transmission sites. This information was used to calculate an emission factor of 67,599 scfy/device. The above emission factors, a methane content of 93.4% and proportions of each of these controllers at transmission sites was used to calculate a "generic" emission factor of 162,197 scfy/device of methane for pneumatic controllers at transmission stations.

Processing

The site visit information from nine natural gas processing plants found that plants used compressed air to operate the majority of pneumatic controllers at the plants. Only one of the plants used natural gas-powered continuous bleed controllers, and five had natural gas-driven pneumatic controllers for the isolation valves on the main pipeline emergency shutdown system or isolation valves used for maintenance. The same type of pneumatic controllers used in the transmission sector are used at natural gas processing sites; therefore, the same emission factors were used to calculate a facility pneumatic emission factor. Using the survey data and the transmission sector pneumatic controller emission factors, the annual methane emissions were determined to be 165 thousand standard cubic feet per facility (Mscfy/facility).

Summary

A summary of the pneumatic controller emission factors, activity factors, and annual methane emission rates estimated by this report are provided in Table 2-2 for the natural gas production, processing and transmission segments. The total methane emissions from pneumatic controllers was estimated to be 45,634 million standard cubic feet per year (MMscfy) or 861,704 metric tons (MT).

---

[8] Shafer Valve Operating Systems. Gas Consumption Calculation Method for Rotary Vane, Gas/Hydraulic Actuators. Technical Bulletin Data, Bulletin GC-2-00394. March 1994.

[9] Personal correspondence with Belva Short of Limitorque Corporation, Lynchburg, VA, April 5, 1994.

Table 2-2 GRI Nationwide Pneumatic Controller Methane Emissions in the United States (1992 Base Year)

Natural gas Segment	Methane Emission Factor	Activity Factor	Annual Methane Emission Rate (MMscf/yr)	Annual Methane Emission Rate (MT)
Production	125,925 scfy/device	249,111 controllers	31,369	592,349
Processing	165,000 scfy/facility	726 facilities	120	2,262
Transmission	162,197 scfy/device	87,206 controllers	14,145	267,093
Total			45,634	861,704

2.1.2 Estimates of Methane Emissions from the U.S. Oil Industry (ICF Consulting, 1999)

ICF Consulting (ICF Consulting, 1999) prepared a report for the EPA that estimated methane emissions from crude oil production, transportation and refining, identified potential methane mitigation techniques and provided an analysis of the economics of reducing methane emissions. The report estimated that 97% of the annual methane emissions occur during crude oil production (59.1 billion cubic feet, Bcf or 1,116,000 MT) in 1995. The transportation and refining sectors generate 0.3 Bcf (5,700 MT) and 1.3 Bcf (24,500 MT) of the annual methane emissions, respectively. The annual methane emissions were estimated using methane emission factors and activity factors to calculate the annual methane emissions from the oil industry.

In the production segment, annual vented methane emissions from 13 sources account for 91% (53.8 Bcf or 1,016,000 MT) of the total 1995 methane emissions from crude oil production. Two of these sources: high bleed pneumatic controllers and low bleed pneumatic controllers account for 37% (19.9 Bcf or 376,000 MT) and 7% (3.7 Bcf or 69,900 MT) of the annual vented methane emissions, respectively.

The high bleed pneumatic controller methane emissions were calculated using an emission factor of 345 scfd (GRI/EPA, 1996c). The activity factor for high bleed pneumatic controllers was determined to be 157,581 and assumes that tank batteries with heater treaters have four pneumatic controllers (three level controllers and one pressure controller). Tank batteries without heater treaters were assumed to have three pneumatic controllers. In addition, it was assumed that 35% of the total pneumatic controllers were high bleed, which is based on the percentage of continuous bleed pneumatic controllers determined in the GRI/EPA study (GRI/EPA, 1996c).

The low bleed pneumatic controller methane emission factor was estimated to be 10% of the high bleed methane emission factor or 35 scfd.[10] The activity factor for low bleed controllers was calculated to be 292,650 controllers and was determined using the assumption that 65% of the total pneumatic controllers are intermittent bleed (GRI/EPA, 1996c), which this report assumed to be low bleed pneumatic devices.

No methane emissions from pneumatic controllers were estimated in this report for the transportation and refining segments of the oil industry.

2.1.3   Inventory of U.S. Greenhouse Gas Emissions and Sinks: 1990-2012 (U.S. EPA, 2014)

The EPA leads the development of the annual Inventory of U.S. Greenhouse Gas Emissions and Sinks (GHG Inventory). This report tracks total U.S. greenhouse gas (GHG) emissions and removals by source and by economic sector over a time series, beginning with 1990.

The U.S. submits the GHG Inventory to the United Nations Framework Convention on Climate Change (UNFCCC) as an annual reporting requirement. The GHG Inventory includes estimates of methane and carbon dioxide for natural gas systems (production through distribution) and petroleum systems (production through refining).

---

[10] EPA Natural Gas STAR default value for low bleed pneumatic controllers.

13

Table 2-3 summarizes the 2014 GHG Inventory's (published in 2014; containing emissions data for 1990-2012) estimates of 2012 national methane emissions from pneumatic controllers in the natural gas production, processing, transmission and storage segments and the petroleum production segment. Where presented in the GHG Inventory, the table includes potential emissions (i.e., emissions that would be released in the absence of controls), emission reductions and net emissions. For pneumatic controllers, the emission reductions reported to the Natural Gas STAR program are deducted from potential emission to calculate net emissions. In future years, the GHG Inventory will also account for regulatory reductions impacting emissions from pneumatic controllers that result from subpart OOOO.

Table 2-3. Summary of GHG Inventory 2012 Nationwide Emissions from Pneumatic Controllers

Industry Segment	Potential CH$_4$ Emissions (MT)	CH$_4$ Emission Reductions (MT)	Net CH$_4$ Emissions (MT)
Natural gas and petroleum production[a]	1,642,622	873,100	769,522
Natural gas processing	1,923	[b]	[b]
Natural gas transmission and storage	263,561	14,078	249,483

[a] In the GHG Inventory, all Natural Gas STAR reductions for pneumatic devices are removed from the natural gas systems estimate. As some of these reductions likely occur in petroleum systems, a combined number for production segment pneumatic devices in natural gas and petroleum systems is presented here.
[b] The GHG Inventory does not include a specific emission reduction for pneumatic controllers in the natural gas processing sector resulting from the Natural Gas STAR program although it is likely non-zero.

The GHG Inventory data estimates that pneumatic controller emissions are 13% of overall methane emissions from the oil and natural gas sectors. The following sections provide greater detail on the estimates given in Table 2-3.

2.1.3.1 Natural gas and petroleum production industry segment

Table 2-4 shows the 2014 GHG Inventory's estimates of 2012 methane emissions from pneumatic controllers in the natural gas and petroleum production industry segment. The table

14

presents the population of pneumatic controllers, methane emission factors, potential methane emissions, and the estimated national total of pneumatic controllers and potential methane emissions. The natural gas production data are broken down by the Energy Information Agency's (EIA's) National Energy Modeling System (NEMS) regions. The table also presents the national total of methane emission reductions compiled from Natural Gas STAR reports and the resulting estimated national net methane emissions from pneumatic controllers.

Table 2-4. Estimated 2012 National and Regional Methane Emissions from Pneumatic Controllers in the Natural Gas and Petroleum Production Segment

NEMS Region	Population of Pneumatic Controllers[a]	CH$_4$ Potential Emission Factor (scfd/device)[a]	CH$_4$ Emissions (MT)
Potential Emissions-Natural Gas Systems			
North East	77,261	373	202,696
Midcontinent	167,589	362	426,133
Rocky Mountain	122,127	339	291,166
South West	55,095	353	136,534
West Coast	2,098	402	5,933
Gulf Coast	53,436	386	145,057
Total	**477,606**		**1,207,519**
Potential Emissions-Petroleum Systems			
High Bleed	145,179	330	336,692
Low Bleed	269,618	52	98,411
Total	**414,797**		**435,103**
Combined Natural Gas and Petroleum Systems			
Total	**892,403**		**1,642,622**
Voluntary Emission Reductions-Natural Gas and Petroleum			873,100
Net Emissions-Natural Gas and Petroleum[b]			769,522

[a] 1996 GRI/EPA report, extrapolated using ratios relating other factors for which activity data are available.

[b] In the GHG Inventory, all Natural Gas STAR reductions for pneumatic devices are removed from the natural gas systems estimate. As some of these reductions likely occur in petroleum systems, a combined number for production segment pneumatic devices in natural gas and petroleum systems is presented here.

Recent national activity data on pneumatic controllers are not available. To calculate national emissions for these sources for the GHG Inventory, a set of industry activity data drivers was developed and used to update activity data. For the natural gas production segment, pneumatic controllers were estimated each year by applying a regional factor for the number of pneumatic controllers per well to annual regional data on gas well population. These factors ranged from 0.5 to 1.6 pneumatic controllers per well. For the petroleum production segment, pneumatic controllers were estimated each year by applying a factor for the number of pneumatic controllers per heater/treater (4), and pneumatic controller per battery without a heater/treater (3).

The basis for the GHG Inventory's potential methane emission factors for pneumatic controllers in the natural gas and petroleum production industry segment is the 1996 GRI/EPA report. The factor for natural gas systems represents a mix of the average emissions from continuous bleed and intermittent natural gas-driven pneumatic controllers in the 1996 GRI/EPA report. The region-specific factors are developed using the GRI/EPA factor and regional gas composition data. For petroleum systems, it was then assumed that 65% of pneumatic controllers in the petroleum production segment are low bleed pneumatic controllers, and 35% of controllers are high bleed. The GRI/EPA factors for low and high bleed controllers are applied to these populations

According to the GHG Inventory, the 1996 GRI/EPA report "still represents the best available [emissions] data in many cases, [but] using these emission factors alone to represent actual emissions without adjusting for emissions controls would in many cases overestimate emissions. For this reason, 'potential emissions' are calculated using the [1996 GRI/EPA report] data, and then current data on voluntary and regulatory emission reduction activities are deducted to calculate actual emissions."

In the case of pneumatic controllers in the natural gas production industry segment, the GHG Inventory reduces the calculated potential emissions using voluntary emission reductions reported by industry partners to the Natural Gas STAR Program. The reductions undergo quality assurance and quality control checks to identify errors, inconsistencies, or irregular data before being incorporated into the GHG Inventory. Future inventories are expected to reflect the subpart OOOO requirements for pneumatic controllers as they are implemented.

2.1.3.2 Natural gas processing industry segment

Table 2-5 shows the 2014 GHG Inventory's estimates of 2012 methane emissions from pneumatic controllers in the natural gas processing industry segment.

Table 2-5. Estimated 2012 National Methane Emissions from Pneumatic Controllers in the Natural Gas Processing Segment

Activity Factor	CH$_4$ Potential Emission Factor (scfy/plant)[b]	CH$_4$ Potential Emissions (MT)	CH$_4$ Emission Reductions (MT)	Net CH$_4$ Emissions (MT)
606 gas plants[a]	164,721	1,923	[c]	[c]

[a] *Oil and Gas Journal*, with available 2012 activity data.
[b] 1996 GRI/EPA report.
[c] Although voluntary Natural Gas STAR emission reductions are reported for this industry segment in the aggregate, no value is given specifically for pneumatic controllers.

The basis for the GHG Inventory's potential methane emission factors for pneumatic controllers in the natural gas processing segment is the 1996 GRI/EPA report. This potential emission factor is expressed in terms of standard cubic feet per year per processing plant (scfy/plant). The associated activity factor is the number of U.S. gas plants, which comes from the *Oil and Gas Journal*.

The GHG Inventory does not report emissions reductions specific to pneumatic controllers in this industry segment and, thus, there is no reported net emissions figure.

2.1.3.3 Natural gas transmission and storage segment

Table 2-6 shows the 2014 GHG Inventory's estimates of 2012 methane emissions from pneumatic controllers in the natural gas transmission and storage industry segment.

Table 2-6. Estimated 2012 National Methane Emissions from Pneumatic Controllers in the Natural Gas Transmission and Storage Segment

Subsegment	Activity Factor (# of controllers)	CH₄ Potential Emission Factor (scfy/device)	CH₄ Potential Emissions (MT)	CH₄ Emission Reductions (MT)	Net CH₄ Emissions (MT)
Transmission	70,827	162,197[a]	221,257		
Storage	13,542	162,197[a]	42,304		
Total	**84,369**		**263,561**	**-14,078[b]**	**249,483**

[a] 1996 GRI/EPA report.
[b] Voluntary Natural Gas STAR emission reductions are reported for all pneumatic controllers in this industry segment, not split out by transmission and storage.

The basis for the GHG Inventory's potential methane emission factors for pneumatic controllers in the natural gas transmission and storage segment is the 1996 GRI/EPA report. In this case, the potential emission factor is expressed in terms of scfy/device. The associated activity factor is the number of pneumatic controllers. For transmission, the number of pneumatic controllers is calculated based on transmission pipeline length. For storage, the number of pneumatic controllers is calculated based on number of compressor stations in the storage segment.

The 2014 GHG Inventory includes voluntary emission reductions reported by industry partners to the Natural Gas STAR Program for pneumatic controllers in the natural gas transmission and storage industry segment.

2.1.4   Greenhouse Gas Reporting Program (U.S. EPA, 2013)

In October 2013, the EPA released 2012 GHG data for Petroleum and Natural Gas Systems collected under the Greenhouse Gas Reporting Program (GHGRP). The GHGRP, which was required by Congress in the FY2008 Consolidated Appropriations Act, requires facilities to report data from large emission sources across a range of industry sectors, as well as suppliers of certain GHGs and products that would emit GHGs if released or combusted.

When reviewing this data and comparing it to other data sets or published literature, it is important to understand the GHGRP reporting requirements and the impacts of these requirements on the reported data. The GHGRP covers a subset of national emissions from Petroleum and Natural Gas Systems; a facility in the Petroleum and Natural Gas Systems source category is required to submit annual reports if total emissions are 25,000 MT of CO_2 equivalent (MT CO_2e) or more. Facilities use uniform methods prescribed by the EPA to calculate GHG emissions, such as direct measurement, engineering calculations, or emission factors derived from direct measurement. In some cases, facilities have a choice of calculation methods for an emission source.

The GHGRP addresses petroleum and natural gas systems with implementing regulations at 40 CFR part 98, subpart W. The rules define three segments of the oil and natural gas industry sector that are required to report GHG emissions from pneumatic controllers: (1) onshore petroleum and natural gas production, (2) onshore natural gas transmission compression, and (3) underground natural gas storage. Facilities calculate emissions from pneumatic controllers by determining the number of each type of controller at the facility and applying emission factors. In the petroleum and natural gas production segment, facilities must apply facility-specific gas composition factors for methane and CO_2. In the natural gas transmission and storage segments, default gas composition factors are used. Subpart W emission factors for pneumatic controllers are located at 40 CFR Part 98, subpart W, Table W-1A (Onshore Petroleum and Natural Gas Production), Table W-3 (Onshore Natural Gas Transmission Compression), and Table W-4 (Underground Natural Gas Storage). These emission factors are based on the 2009 document *Compendium of Greenhouse Gas Emissions Methodologies for the Oil and Gas Industry* published by the American Petroleum Institute (API), which in turn is based on the 1996 GRI/EPA report.

Table 2-7 shows the number of reporting facilities[11] in each of the three industry segments, along with reported pneumatic controller methane emissions.

Table 2-7. Facilities and Reported Emissions from Pneumatic Controllers, 2012

Segment	Number of Reporting Facilities	Reported Methane Emissions (MT)[a]
Petroleum and NG Production	417	861,224
Transmission	330	7,582
Storage	38	4,493

[a] The reported methane MT CO_2e emissions were converted to methane emissions in MT by dividing by a global warming potential (GWP) of methane (21).

2.1.5   Measurements of Methane Emissions at Natural Gas Production Sites in the United States (Allen et al., 2013)

A study completed by multiple academic institutions and consulting firms was conducted to gather methane emissions data at onshore natural gas sites in the U.S. and compare those emission estimates to the 2011 estimates reported in the 2013 GHG Inventory. The sources or operations tested included 305 pneumatic controllers located at 150 distinct natural gas production sites in four production regions (Appalachian, Gulf Coast, Midcontinent, and Rocky Mountain).

Testing was carried out using a Hi-Flow Sampler, which is a portable, battery-powered instrument designed to determine the rate of gas leakage around various pipe fittings, valve packings and compressor seals found in natural gas production, transmission, storage and processing facilities. To allow the quantity of methane to be separated out from other chemical

---

[11] In general, a "facility" for purposes of the GHGRP means all co-located emission sources that are commonly owned or operated. However, the GHGRP has developed a specialized facility definition for onshore production. For onshore production, the "facility" includes all emissions associated with wells owned or operated by a single company in a specific hydrocarbon-producing basin (as defined by the geologic provinces published by the American Association of Petroleum Geologists).

species, gas composition data were collected for each natural gas production site, typically provided by the site owner. The 305 sampled pneumatic controllers represented an estimated 41% of all the controllers associated with the wells that were sampled. The sampling time for each controller was not specified in the study. Table 2-8 shows the emission rates determined by the testing.

Table 2-8. Pneumatic Controller Methane Emission Rates Reported in the Allen Study

	Methane Emissions per Pneumatic Controller				
	Appalachian	**Gulf Coast**	**Midcontinent**	**Rocky Mtn.**	**Total**
Number sampled[a]	133	106	51	15	305
Emissions rate (scf methane/min/device)[b]	0.126 ± 0.043	0.268 ± 0.068	0.157 ± 0.083	0.015 ± 0.016	0.175 ± 0.034
Emissions rate (scf whole gas/min/device, based on site-specific gas composition)[b]	0.130 ± 0.044	0.289 ± 0.071	0.172 ± 0.086	0.021 ± 0.022	0.187 ± 0.036

[a] Intermittent and low bleed controllers are included in the total; no high bleed controllers were reported by companies providing controller type information
[b] Uncertainty characterizes the variability in the mean of the data set, rather than an instrumental uncertainty in a single measurement

The Allen study reports that the average whole gas emission rate was 11.2 scfh per pneumatic controller for the tested population, which consisted of a mix of intermittent and low bleed controllers. No high bleed controllers were reported by the companies that provided controller type information. The study also reports whole gas emission factors of 5.1 scfh for low bleed controllers and 17.4 scfh for intermittent controllers. These emission factors are based on measured emissions at the 24 sites where the site operators reported only low bleed controllers and the 55 sites reporting only intermittent controllers, where potential misidentification of controller type is less likely to be a confounding factor.

The study notes that there is significant geographical variability in the emissions rate from pneumatic controllers between production regions. Emissions per controller from the Gulf Coast are highest and are statistically different than emissions from controllers in the Rocky Mountain and Appalachian regions. The difference in average values is more than a factor of 10 between Rocky Mountain and Gulf Coast regions. The study provided the following discussion of these differences:

> Some of the regional differences in emissions may be explained by differences in practices for utilizing low bleed and intermittent controllers. For example, new controllers installed after February 1, 2009 in regions in Colorado that do not meet ozone standards, where most of the Rocky Mountain controllers were sampled, are required to be low bleed (or equivalent) where technically feasible (Colorado Air Regulation XVIII.C.1; XVIII.C.2; technical feasibility criterion under review as this is being written). However, observed differences in emission rates between intermittent and low bleed devices (roughly a factor of 3) are not sufficient to explain all of the regional differences. A number of additional hypotheses were examined to attempt to explain the differences in emissions. For datasets consisting entirely of intermittent or entirely of low-bleed devices, the volume of oil produced was not a good predictor of emissions. Wellhead and separator pressure were also not good predictors of emissions. The definition of low-bleed controllers may be [an] issue, however. All low bleed devices are required to have emissions below 6 scf/hr (0.1 scf/m), but there is not currently a clear definition of which specific controller designs should be classified as low bleed and reporting practices among companies can vary. Other possibilities for explaining the low-bleed emission rates observed in this work, that have not yet been investigated, but that may be pursued in follow-up work, include operating practices for the use of the controllers.

The study estimated 2011 national methane emissions from pneumatic controllers in the natural gas production industry segment at 570,000 MT (with a range of 510,000 – 812,000 MT based on the 95% confidence bounds of the emission factor) using the same number of controllers (447,379) used in the 2013 GHG Inventory for 2011. This estimate was computed using a regionally weighted emission factor of 67,400 scfy methane/device.

2.1.6 Determining Bleed Rates for Pneumatic Devices in British Columbia (Prasino Group 2013)

A study completed by the Prasino Group was conducted to determine the average bleed rate of pneumatic controllers when operating under field conditions in British Columbia (BC). Bleed rates were sampled from pneumatic controllers using a positive displacement bellows meter at upstream oil and gas facilities across a variety of producing fields in the Fort St. John, BC and surrounding areas. For this study, bleed rate was defined as "the amount of fuel gas released to the atmosphere per hour," including both continuous bleed (where applicable) and emissions during activation. The study centered on high bleed controllers, including both continuous bleed and intermittent controllers with emissions greater than 0.17 cubic meters per hour (m^3/hr) (e.g, > 6 scfh).[12] The study aimed to identify the most common high bleed pneumatic controllers in the field and test emissions from at least 30 units of each model. In identifying controllers to test, the study used a manufacturer-supplied emission rate of 0.119 m^3/hr as a cutoff to explore whether some models identified by manufacturers as low bleed perform at that level in the field.

Field measurements were carried out with a Calscan Hawk 9000 Vent Gas Meter, which uses a positive displacement diaphragm meter that detects flow rates down to zero, and can also effectively measure any type of vent gas (methane, air, or propane). (A few sampled devices ran on air at large sites using compressed air, or propane at sour sites using compressed propane; such samples were corrected using a density ratio to equivalent natural gas emissions rates.) This device uses "a precision pressure sensor, an external temperature probe, and industry standard gas flow measurement algorithms to accurately measure the gas rates and correct for pressure and temperature differences." The report notes that metering a device can affect the operation of the device when hooked up due to back pressure, adding that it is possible that certain controllers did not produce enough pressure when hooked up to overcome the back pressure, resulting in a

---

[12] This definition of "high bleed" is slightly different than the definition presented in Section 1.1.1 of this paper, because "intermittent bleed controllers" are included as "high bleed controllers" if their emissions are above the specified threshold. The definition presented in Section 1.1.1 places "intermittent bleed controllers" in their own category.

zero reading. The sample time for each controller was 30 minutes, so there was variability in the number of actuation events captured for each controller depending on operating conditions.

In addition to emission factors for individual models of pneumatic controllers, the study generated emission factors for "generic high bleed controllers" and "generic high bleed intermittent controllers." The study also included a regression analysis of the relationship between bleed rate and the pressure of the supply gas routed to the controller. Based on the analysis, the study found that the positive relationship between these parameters was strong enough to recommend use of a supply pressure coefficient to calculate the bleed rate for several controller models and for generic controllers. The generic emission factors and supply pressure coefficients are shown in Table 2-9.

Table 2-9. Generic Natural Gas Emission Factors and Supply Pressure Coefficients for High Bleed Pneumatic Controllers

Type of Pneumatic Controller	Average Bleed Rate (m³/hr)[a]	Average Bleed Rate (scfh)[b]	Coefficient Related to Supply Pressure[c]
Generic High Bleed Controller	0.2605	9.199	0.0012
Generic High Bleed Intermittent Controller	0.2476	8.744	0.0012

[a] "Bleed rate" defined to include actuation emissions as well as continuous bleed.
[b] Calculated.
[c] Supply pressure apparently in kPa, although not clearly stated in the report.

Based on what it termed a "positive correlation," the Prasino study recommended the use of the supply pressure coefficients above for calculating emission rates for generic high bleed controllers and generic high bleed intermittent controllers. It should be noted that the coefficients of determination (R^2 values) for these supply pressure coefficients are 0.41 and 0.35 for high bleed and high bleed intermittent controllers, respectively.

2.1.7 Air Pollutant Emissions from the Development, Production, and Processing of Marcellus Shale Natural Gas (Roy et al., 2014)

A study by the Center for Atmospheric Particle Studies at Carnegie Mellon University was conducted to develop an emission inventory for the development, production, and processing of natural gas in the Marcellus Shale region for 2009 and 2020. (Note: The focus of this white paper is current emissions, therefore, the 2020 projections are not discussed further.) The inventory includes estimates for emissions of nitrogen oxides, VOC, and particulate matter less than 2.5 micrometers in diameter from major activities, including VOC emissions from pneumatic controllers associated with "wet" and "dry" gas wells. The study estimated VOC emissions from pneumatic controllers associated with Marcellus Shale natural gas wells to be on the order of 10 tons/day in 2009.

This study developed these emissions estimates by estimating the number of wet and dry wells in the region and establishing per-well emission factors for 2009. The per-well emission factors are shown in Table 2-10.

Table 2-10. Per-Well VOC Emissions from Pneumatic Controllers in 2009 (95% confidence interval)

Type of Well	VOC Emissions, 2009 (tons/producing well)
Dry Gas	0.5 (0.08 – 0.8)
Wet Gas	3.5 (2.4 – 4.4)

The per-well emission factors were based on assumptions regarding the type, number, and emission factors for pneumatic controllers associated with each natural gas well, which were drawn primarily from a 2008 ENVIRON report.[13] Table 2-11 shows these assumptions.

---

[13] Bar-Ilan, Amnon et al., ENVIRON International Corporation. *Recommendations for Improvement to the CENRAP States' Oil and Gas Emissions Inventories*. Prepared for the Central States Regional Air Partnership. November 13, 2008. This report also includes emission factors for positioners (15.2 scfh) and transducers

Table 2-11. Assumed Type, Number, and Emission Factors for Pneumatic Controllers Associated with Each Natural Gas Well

Type of Device	Number of Controllers	Emission Factor (scfh)[a]
Liquid Level Controller	2	31 (2009)
Pressure Controller	1	17 (2009)

[a] 2009 emission factors are from the 2008 ENVIRON report.

The emission factors from this study and the ENVIRON report are not comparable to the emission factors discussed above because they are provided for different classifications of pneumatic controllers. In addition, these emission factors differ from those discussed previously in that they are based on bleed rates provided by manufacturers rather than measured emissions.

2.1.8 Economic Analysis of Methane Emission Reduction Opportunities in the U.S. Onshore Oil and Natural Gas Industries (ICF, 2014)

The Environmental Defense Fund (EDF) commissioned ICF International (ICF) to conduct an economic analysis of methane emission reduction opportunities from the oil and natural gas industry to identify the most cost-effective approach to reduce methane emissions from the industry. The study projects the estimated growth of methane emissions through 2018 and focuses its analysis on 22 methane emission sources in the oil and natural gas industry (referred to as the targeted emission sources). These targeted emission sources represent 80% of their projected 2018 methane emissions from onshore oil and gas industry sources. Pneumatic devices are several of the 22 emission sources that are included in the study and include: high bleed pneumatic controllers, intermittent bleed pneumatic controllers, Kimray pumps, intermittent bleed pneumatic controllers – dump valves, and chemical injection pumps. The

---

(13.6 scfh). The emission factors in the report "were obtained from data gathered as part of the EPA's Natural Gas STAR program." Examination of Natural Gas STAR program materials clearly shows that these emission factors were derived from the manufacturer-supplied natural gas bleed rates for high bleed pneumatic controllers listed in Appendix A to *Options for Reducing Methane Emissions from Pneumatic Devices in the Natural Gas Industry*. The ENVIRON report also includes the number of controllers of each type associated with each gas well, said to be drawn from survey data in the CENRAP states. The numbers for liquid level controllers and pressure controllers are reflected in Table 2-15; the report found zero positioners and transducers per well.

methodology that was used for this analysis is based on the 2013 GHG Inventory and uses data from the GHGRP and the University of Texas/EDF gas production measurement study (Allen et al., 2013).

The study relied on the 2013 GHG Inventory for 1990-2011 for methane emissions data for the oil and natural gas sector. These emissions data were revised to include updated information from the GHG Inventory and the *Measurements of Methane Emissions at Natural Gas Production Sites in the United States* study (Allen et al., 2013). The revised 2011 baseline methane emissions estimate was used as the basis for projecting onshore methane emissions to 2018. (Note: The focus of this white paper is current emissions, therefore, the 2018 projections are not discussed further.)

The study used the 2013 GHG Inventory estimates for 2011 to develop new activity and emission factors for pneumatic controllers. The count of pneumatic controllers was calculated using the well counts and assuming 0.94 pneumatic controllers per well. The study did find that there are an additional 8.6 pneumatic controllers per gathering/boosting station that were not accounted for in the 2013 GHG Inventory. The study also used emission factors from subpart W, which reported pneumatic controllers in three categories: low bleed, intermittent bleed and high bleed controllers. To break out the number of pneumatic controllers in each of these categories, the emission data from subpart W were analyzed, and the study determined that the percentage of pneumatic controllers were 10% high bleed, 50% intermittent bleed and 40% low bleed. These percentages were applied to the pneumatic controller counts and the respective emission factor was used to calculate the emissions from these controllers. Intermittent pneumatic controllers were further segregated into two categories: dump valves and non-dump valve intermittent controllers. The dump valves represent intermittent controllers that do not continuously bleed and only emit during actuation. The study estimated that 75% of the total intermittent pneumatic controllers were dump valves. Based on the subpart W data and the assumptions above, the study used the following emission factors for each of the controllers: 320 Mcf/yr/device for high bleed, 120 Mcf/yr/device for non-dump intermittent, 20 Mcf/yr/device for dump intermittent and 11 Mcf/yr/device for low bleed pneumatic controllers. Using these factors, the study estimated an increase of 41% (26 Bcf or 491,000 MT) in methane emissions in comparison to the 2013 GHG Inventory.

Further information included in this study on the replacement of high bleed and intermittent bleed pneumatic controllers with low bleed pneumatic controllers, and the replacement of pneumatic pumps with electric pumps as mitigation or emission reduction techniques, methane control costs, and their estimates for the potential for VOC emissions co-control benefits from the replacement of these pneumatic controllers are presented in Section 3 of this document.

2.2   Discussion of Data Sources for Pneumatic Pumps

Many of the data sources for pneumatic pumps are the same as those for pneumatic controllers, therefore, the overall descriptions of these data sources are not repeated in this section and only the information relevant to pneumatic pumps is discussed.

2.2.1   Methane Emissions from the Natural Gas Industry (GRI/EPA, 1996c) (GRI/EPA, 1996e)

The methane emission estimates for pneumatic pumps are separated into two categories for the GRI/EPA reports; chemical injection pumps (GRI/EPA, 1996d) and gas-assisted glycol pumps (GRI/EPA, 1996f). A summary of each of these reports and the methane calculation methodologies are provided in the following sections.

2.2.1.1 Methane Emissions from the Natural Gas Industry – Chemical Injection Pumps (GRI/EPA, 1996c)

This report estimates emissions from two types of pumps that the oil and natural gas industry uses for chemical injection into process streams: piston pumps and diaphragm pumps. Four sources of information were used to develop an emission factor for chemical injection pumps: a study by the CPA[14], data collected from site visits, literature data for methane composition, and data from pump manufacturers.

The CPA study provided natural gas emissions from five diaphragm chemical injection pumps using the "bagging" method. This method involves enclosing the pump and measuring the flow rate and concentration of the natural gas emissions from the pump. The measurements from this study reported natural gas emissions ranging from 254 to 499 standard cubic feet per day per pump (scfd/pump) with an average of 334 scfd/pump.

Data from site visits included: the total number of chemical injection pumps for a particular site, number of chemical injection pumps used in natural gas production, the energy source for the pump (e.g., natural gas, instrument air, electricity), frequency of operation (e.g., pumping rate in strokes per minute), number of pumps that are active or idle, pump operation schedule, size of the unit (e.g., volume displacement of the motive chamber), manufacturer and model number of the pump, and supply gas pressure. Table 2-12 provides a summary of the site visit data. The methane emission factor in Table 2-12 was calculated using a methane composition of 78.8% (GRI/EPA, 1996d).

Table 2-12. Summary of Chemical Injection Pump Site Visit Data

Chemical Injection Pump Data	All Data		Natural Gas Industry Data	
	Piston Pumps	Diaphragm Pumps	Piston Pumps	Diaphragm Pumps
Percent of Total Pumps	49.8 ± 38%	50.2 ± 38%	4.5 ± 678%	95.5 ± 32%
Pump Actuation Rate (strokes/min)	26.32 ± 29%	13.64 ± 49%	3.57 ± 42%	14.75 ± 61%
Number of Pump Actuation Measurements	32	8	15	5
Number of Sites with	7	5	2	4

---

[14] Canadian Petroleum Association. A Detailed Inventory of CH_4 and VOC Emissions from Upstream Oil and Gas Operations in Alberta, March 1992.

Chemical Injection Pump Data	All Data		Natural Gas Industry Data	
	Piston Pumps	Diaphragm Pumps	Piston Pumps	Diaphragm Pumps
Pump Actuation Measurements				
Percent of Pumps Operating	44.6 ± 62%	40.0 ± 52%	77.5 ± 148%	58.0 ± 39%
Number of Sites with Pumps Operating	7	10	4	6
Methane Emissions Factor (scfd/pump)	248 ± 83%		668 ± 88%	

Manufacturers' data and the CPA data were used to determine the volume of gas released per pump stroke. This was done by using the natural gas usage data (amount of natural gas required to pump one gallon of liquid), stroke length, and stoke diameter to calculate volume of natural gas per pump stoke. For diaphragm pumps, the average natural gas usage was calculated to be 0.0719 standard cubic feet per stroke (scf/stroke). The piston pump average natural gas usage was calculated to be 0.0037 scf/stroke. These averages were then used to determine the emission factor for each of the pump types by multiplying the average frequency (strokes per day) by the operating time percentage. Note that the report uses the "all data" frequency and operating percentage to calculate the emission factor for each type of pump. The emission factor for diaphragm pumps was calculated to be 446 scfd/pump and the emission factor for piston pumps was calculated to be 48.9 scfd/pump.

The percentage of piston and diaphragm pumps and their respective emission factors were then used to calculate an average emission factor for chemical injection pumps. The average emission factor was determined to be 248 scfd/pump. The 1992 national emissions were then calculated using the average chemical injection pump emission factor (248 scfd/pump) and the activity factor for chemical injection pumps of 16,971 (GRI/EPA, 1996a). The resulting 1992 national emissions from chemical injection pumps for the natural gas production segment was calculated to be 1,536 MMscf/yr (29,008 MT).

2.2.1.2 Methane Emissions from the Natural Gas Industry – Gas-Assisted Glycol Pumps
(GRI/EPA, 1996e)

For many glycol dehydrators in the natural gas industry, small gas-assisted pumps are used to circulate the glycol. These pumps use energy from the high-pressure rich glycol/gas mixture leaving the absorber to pump the low-pressure lean glycol back to the absorber. Natural gas is entrained in the rich glycol stream feeding the pump and is discharged from the pump at a lower pressure to the regenerator. If the glycol unit has a flash tank, most of the natural gas in the low-pressure stream can be recovered and used as a fuel or stripping gas. If the natural gas from the pump is used as a stripping gas, or if there is no flash tank, all of the pump exhaust gas will be vented through the regenerator's atmospheric vent stack (GRI/EPA, 1996e).

Methane emissions from these gas-assisted pumps were calculated using technical information from Kimray, a manufacturer of gas-assisted pumps. No direct measurements of pump gas usage were used in the calculations. Kimray reported that the natural gas usage ranges from 0.081 actual cubic feet per gallon of glycol pumped (acf/gal) for high-pressure pumps (>400 psig) to 0.130 acf/gal for low-pressure pumps (< 400 psig). These values convert to 3.73 standard cubic feet per gallon (scf/gal) at an operating pressure of 800 psig and 83 mole percent methane for high-pressure pumps and 2.31 scf/gal at an operating pressure of 300 psig and 83 mole percent methane for low-pressure pumps.

The gas usage rates were then converted to an amount of natural gas treated by assuming a typical high-pressure dehydrator would remove 53 pounds of water per million cubic feet of gas (lbs/MMscf), and a typical low-pressure dehydrator would remove 127 lbs/MMscf. The design glycol-to-gas ratio was assumed to be three gallons of glycol per pound of water removed and an overcirculation ratio of 2.1 was used to determine the emission factors for the pumps for the natural gas production segment. Using these factors and the fraction of dehydrators without flash tanks (0.735) and the fraction of dehydrators without combustion vent controls (0.988), the emission factor for the gas-assisted pumps in the natural gas production segment were calculated to be 904.5 standard cubic feet of methane per million standard cubic feet of natural gas treated (scf/MMscf) for high-pressure pumps, and 1342.2 scf/MMscf for low-pressure pumps. The final emission factor for methane from an average gas-assisted glycol pump was determined assuming that 80% of these pumps are high-pressure and 20% are low-pressure. The average emission factor was calculated to be 992.0 scf/MMscf and was used to estimate methane emissions from the natural gas production segment.

For natural gas processing, the study assumed that only high-pressure gas-assisted glycol pumps are used. The emission factor was calculated using the high-pressure pump gas usage (3.73 scf/gal), the design glycol-to-gas ratio (3 gal glycol/lb water), the water removal rate for a high-pressure system (53 lbs/MMcsf), an overcirculation ratio of 1.0, the fraction of dehydrators without flash tanks (0.333) and the fraction of dehydrators without combustion vent controls (0.900). These values were used to calculate a methane emission factor of 177.8 scf/MMscf for gas-assisted pumps for the natural gas processing segment. The natural gas transmission and storage segments do not use gas-assisted glycol pumps.

The 1992 national methane emissions were calculated using data from site surveys to determine the natural gas throughput of dehydrators with gas-assisted pumps. The natural gas throughput of dehydrators with gas-assisted pumps was estimated to be 11.1 trillion standard cubic feet per year (Tscf/yr) for the natural gas production segment and 0.958 Tscf/yr for the natural gas processing segment. The 1992 national methane emissions from gas-assisted pumps were calculated to be 10,962 MMscf/yr (206,989 MT) for the natural gas production segment and 170 MMscf/yr (3,215 MT) for the natural gas processing segment.

2.2.2   Inventory of U.S. Greenhouse Gas Emissions and Sinks: 1990-2012 (U.S. EPA, 2014)

Table 2-13 summarizes the 2014 GHG Inventory estimates of 2012 national methane emissions from pneumatic pumps in the natural gas production and processing segments. (Note: The GHG inventory does not include estimates of emissions from pneumatic pumps in the natural gas transmission and storage segments.) The pneumatic pumps described in the GHG Inventory include: chemical injection pumps and Kimray pumps. The table includes potential emissions, emission reductions and net emissions. For pneumatic pumps, the emission reductions in this report are voluntary reductions through the Natural Gas STAR program. In future years, the GHG Inventory will also account for regulatory reductions that result from subpart OOOO.

Table 2-13. Summary of GHG Inventory 2012 Nationwide Emissions from Pneumatic Pumps

Industry Segment	Potential CH$_4$ Emissions (MT)	CH$_4$ Emission Reductions (MT)	Net CH$_4$ Emissions (MT)
Natural gas production	455,719	2,771	452,948
Petroleum Production	49,973	N/A	
Natural gas processing	5,011	N/A	

The 2014 GHG Inventory data estimates that pneumatic pump emissions are around 16% of overall methane emissions from the natural gas production and processing sectors.

Tables 2-14 and 2-15 show the 2014 GHG Inventory's estimates of 2012 methane emissions from chemical injection pumps and gas-assisted pumps (Kimray pumps) in the natural gas and petroleum production and processing industry segments. The tables present population of chemical injection and Kimray pumps, methane emission factors and potential methane emissions from these devices in each of the EIA's NEMS regions, and the estimated national total of chemical injection pumps, Kimray pumps and potential methane emissions. The activity factors for chemical injection pumps are based on the estimated count of chemical injection pumps in operation. For the production sector, a regional factor for pumps per well (ranging from 0.01 to 0.68) is applied to annual regional well counts to calculate chemical injection pumps each year for natural gas, and for petroleum systems it is estimated that around 20% of wells have pumps (based on 1996 GRI/EPA) and that 25% of pumps use gas. For the production sector, the activity factors for Kimray pumps are based on the total throughput of natural gas multiplied by the fraction of dehydrators using gas-driven pumps (0.9 for the production segment). For the processing segment, the activity factor for Kimray pumps is based the total processing plant throughput multiplied by the fraction of natural gas treated by dehydrators at

33

gas plants (0.5) and then multiplied by the fraction of dehydrators that use a gas-driven pump (0.1 for the processing segment).

Table 2-14. Estimated 2012 National and Regional Methane Emissions from Chemical Injection Pumps in the Natural Gas Production Segment

NEMS Region	Population of Chemical Injection Pumps[a]	CH_4 Potential Emission Factor (scfd/device)[a]	CH_4 Emissions (MT)
Natural Gas Production			
North East	795	268	1,499
Midcontinent	15,343	260	28,045
Rocky Mountain	14,849	244	25,448
South West	2,531	253	4,508
West Coast	1,422	289	2,890
Gulf Coast	2,537	278	4,951
Total Natural Gas	**37,477**		**67,341**
Voluntary Emission Reductions			**-2,771**
Net Emissions-Natural Gas			**64,570**
Petroleum Production	28,702	248	**49,973**

[a] 1996 GRI/EPA report, extrapolated using ratios relating other factors for which activity data are available.

Table 2-15. Estimated 2012 National and Regional Methane Emissions from Kimray Pumps in the Natural Gas Production and Processing Segments

NEMS Region	Total Natural Gas using Kimray Pumps[a]	CH_4 Potential Emission Factor (scfd/MMscf)[a]	CH_4 Emissions (MT)
Natural Gas Production			
North East	6,487,241	1,073	134,073
Midcontinent	4,409,271	1,040	88,322
Rocky Mountain	3,404,114	975	63,934
South West	1,692,957	1,014	33,050
West Coast	85,450	1,157	1,904
Gulf Coast	3,137,482	1,110	67,095
Production Total	**19,216,515**		**388,378**
Natural Gas Processing			

All Regions	1,463,675	178	5,011
Total Potential Emissions			**393,389**

[a] 1996 GRI/EPA report, extrapolated using ratios relating other factors for which activity data are available.
Note: The GHG Inventory did not list any Kimray pumps in the natural gas transmission or distribution sectors.

The basis for the GHG Inventory's potential methane emission factors for pneumatic pumps in the natural gas production and processing segments is the 1996 GRI/EPA report.

The region-specific factors used in the production segment are developed using the GRI/EPA factor and regional gas composition data.

2.2.3   Greenhouse Gas Reporting Program (U.S. EPA, 2013)

The GHGRP addresses petroleum and natural gas systems with implementing regulations at 40 CFR part 98, subpart W. The rule requires facilities in the onshore petroleum and natural gas production segment to report GHG emissions from pneumatic pumps. Facilities calculate emissions from pneumatic pumps by determining the number of pneumatic pumps at the facility and applying an emission factor of 13.3 scf/hour/pump. Facilities also apply a facility-specific gas composition factor for calculating emissions. For 2012, 343 facilities in the onshore petroleum and natural gas production industry segment reported emissions from pneumatic pumps, with total methane emissions of 135,227 metric tons.

2.2.4   Determining Bleed Rates for Pneumatic Devices in British Columbia (Prasino Group 2013)

The study used data from the Canadian Association of Petroleum Producers (2008), Pacific Carbon Trust (2011), Cap-Op Energy's Distributed Energy Efficiency Project Platform (DEEPP) database to compile a list of pneumatic pumps. The study notes that the total number of pneumatic pumps is unknown and the list only comprises a subset of the total population. In total, 184 samples were taken from chemical injection pumps. From the data collected, the study determined the average bleed rate for a piston-type pneumatic pumps to be 0.5917 m^3/hr (approximately 20.9 scfh). For diaphragm-type pneumatic pumps, the bleed rate was calculated to be 1.0542 m^3/hr (approximately 37.2 scfh).

2.2.5   Economic Analysis of Methane Emission Reduction Opportunities in the U.S. Onshore Oil and Natural Gas Industries (ICF, 2014)

The analysis developed by ICF includes an inventory of methane emissions for 2011 using data from the 2013 GHG Inventory and the GHGRP (U.S. EPA, 2013), in addition to data from the EIA and GRI.

For pneumatic chemical injection pumps in the natural gas production segment, the 2011 ICF inventory updated the count of chemical injection pumps to reflect changes made to the well counts and applied the Natural Gas STAR estimated reductions associated with pneumatic pumps. These changes resulted in a 2011 methane estimate of 3 Bcf (56,600 MT) from chemical injection pumps in the natural production segment. Kimray pumps (gas-assisted glycol pumps) were estimated to be 17 Bcf (321,000 MT).

3.0   AVAILABLE PNEUMATIC DEVICE EMISSIONS MITIGATION TECHNIQUES

The following sections describe the different available emissions mitigations techniques that the EPA is aware of for pneumatic controllers and pneumatic pumps. The primary sources of information for mitigations techniques was the EPA's Natural Gas STAR Lessons Learned documents and the ICF economic analysis (ICF, 2014).

3.1   Available Pneumatic Controller Emissions Mitigation Techniques

Several techniques to reduce emissions from pneumatic controllers have been developed over the years. Table 3-1 provides a summary of these techniques for reducing emissions from pneumatic controllers including replacing high bleed controllers with low bleed or zero bleed models, driving controllers with instrument air rather than natural gas, using non-gas-driven controllers, and enhanced maintenance.

Table 3-1. Summary of Alternative Mitigation Techniques for Pneumatic Controllers

Option	Description	Applicability	Costs	Efficacy and Prevalence
Install Zero Bleed Controller in Place of Continuous Bleed Controller (U.S. EPA, 2011a, GE Energy Services, 2012)	Zero bleed controllers are self-contained natural gas-driven devices that vent to the downstream pipeline, not the atmosphere. Provide the same functional control feed, and city gate stations/distribution as continuous bleed controllers, where applicable (U.S. EPA, 2011a, GE Energy Services, 2012).	Applicable only for relatively low-pressure control valves, e.g., in gathering, metering and regulation stations, power plant and industrial applications (U.S. EPA, 2011a).	The EPA does not have information on this technology.	100% emission reduction, where applicable. The EPA does not have information on the prevalence of this technology in the field, however, it is the EPA's understanding that applicability is limited.
Install Low Bleed Controller in Place of High Bleed Controller (U.S. EPA, 2006b)	Low bleed controllers provide the same functional control as a high bleed devices, while emitting less continuous bleed emissions (U.S. EPA, 2006b).	Applicability depends on the function of instrumentation for an individual device and whether the device is a level, pressure, or temperature controller. Not recommended for control of very large valves that require fast and/or precise response to process changes. These are found most frequently on large compressor discharge and bypass pressure controllers (U.S. EPA, 2006b).	Based on information from Natural Gas STAR (U.S. EPA, 2006b) and 2011a): OOOO, low bleed devices cost, on average, around $165 more than high bleed versions. ICF report assumed a cost of $3,000 per replacement based on industry comments (ICF, 2014).	Estimated average reductions (U.S. EPA, 2011a): *Production segment:* 6.6 tpy methane *Transmission:* 3.7 tpy methane The EPA does not have information on the prevalence of this technology in the field.

38

Table 3-1. Summary of Alternative Mitigation Techniques for Pneumatic Controllers

Option	Description	Applicability	Costs	Efficacy and Prevalence
Convert to Instrument Air (U.S. EPA, 2006c)	Compressed air may be substituted for natural gas in pneumatic systems without altering any of the parts of the pneumatic control. In this type of system, atmospheric air is compressed, stored in a tank, filtered and then dried for instrument use. Instrument air conversion requires additional equipment to properly compress and control the pressured air. This equipment includes a compressor, power source, air dehydrator and air storage vessel (U.S. EPA, 2006c).	Most applicable at facilities where there are a high concentration of pneumatic control valves and an operator present. Because the systems are powered by electric compressors, they require a constant source of electrical power or a backup natural gas pneumatic device (U.S. EPA, 2006c).	System costs are dependent on size of compressor, power supply needs, labor and other equipment (U.S. EPA, 2006c). A cost analysis is provided in Section 3.1.3 below.	100% emission reduction, where applicable. There are secondary emissions associated with electrical power generation. The EPA does not have information on the prevalence of this technology in the field.
Mechanical and Solar-Powered Systems in Place of Bleed Controller (U.S. EPA, 2006a, U.S. EPA, 2006b)	Mechanical controls operate using a simple design comprised of levers, hand wheels, springs and flow channels. The most common mechanical control device is the liquid-level float to the drain valve position with mechanical linkages. Electricity or small electrical motors (including solar-powered) have been used to operate valves. Solar control systems are driven by solar power cells that actuate mechanical devices using electric power. As such, solar cells require some type of backup power or storage to ensure reliability (U.S. EPA, 2006a).	Application of mechanical controls is limited because the control must be located in close proximity to the process measurement. Mechanical systems are also incapable of handling larger flow fluctuations. Electric-powered valves are only reliable with a constant supply of electricity (U.S. EPA, 2006a).	Depending on supply of power, costs can range from below $1,000 to $10,000 for entire systems (U.S. EPA, 2006a).	100% emission reduction, where applicable. The EPA does not have information on the prevalence of this technology in the field.

39

Table 3-1. Summary of Alternative Mitigation Techniques for Pneumatic Controllers

Option	Description	Applicability	Costs	Efficacy and Prevalence
Enhanced Maintenance (U.S. EPA, 2006a)	Instrumentation in poor condition typically bleeds 5 to 10 scfh more than representative conditions due to worn seals, gaskets, diaphragms; nozzle corrosion or wear; or loose control tube fittings. This may not impact operations but does increase emissions. Proper methods of maintaining a device are highly variable (U.S. EPA, 2006a).	Enhanced maintenance to repair and maintain pneumatic controllers periodically can reduce emissions at many controllers (U.S. EPA, 2006a).	Variable based on labor, time, and fuel required to travel to many remote locations.	Natural gas emission reductions of 5 to 10 scfh (U.S. EPA, 2006a). The EPA does not have information on the prevalence of this practice in the field.

The mitigation techniques summarized in Table 3-1 are discussed in more detail in the following sections.

3.1.1   Zero Bleed Pneumatic Controllers

Zero bleed pneumatic controllers are self-contained, "closed loop" natural gas-driven controllers that vent to the downstream pipeline rather than to the atmosphere (U.S. EPA, 2011a). These closed loop devices are considered to emit no natural gas to the atmosphere. However, they can be used only in applications with very low pressure and, therefore, may not be suitable to replace continuous bleed pneumatic controllers in many applications. Some applications where they may suitable include gathering, metering and regulation stations, power plant and industrial feed, and city gate stations/distribution (U.S. EPA, 2011a). To date, the EPA has not obtained any information on the cost of zero bleed controllers or their prevalence in the field.

3.1.2   Low Bleed Pneumatic Controllers

Description

Low bleed controllers provide similar functional control as high bleed controllers, but have lower continuous bleed emissions. It has been estimated on average that 6.6 tons of methane and 1.8 tons of VOC will be reduced annually in the production segment from installing a low bleed device in place of a high bleed device (U.S. EPA, 2011a). In the transmission segment, the average achievable reductions per device are estimated around 3.7 tons and 0.08 tons for methane and VOC, respectively (U.S. EPA, 2011a). As defined in this white paper, a low bleed controller can emit up to 6 scfh, but this is higher than the expected emissions from the typical low bleed controllers available on the current market.

Applicability

There are certain situations in which replacing and retrofitting are not feasible, such as instances where a minimal response time is needed, cases where large valves require a high bleed rate to actuate, or a safety isolation valve is involved.

Replacing high bleed pneumatic with low bleed controllers is infeasible in situations where a process condition may require a fast or precise control response so that it does not stray too far from the desired set point (U.S. EPA, 2011a). A slower-acting controller could potentially result in damage to equipment and/or become a safety issue. An example of this is on a compressor where pneumatic controllers monitor the suction and discharge pressure and actuate a recycle when one or the other is out of the specified target range. Another scenario where fast and precise control is necessary includes transient (non-steady) situations where a gas flow rate may fluctuate widely or unpredictably (U.S. EPA, 2011a). In this case, a responsive high bleed device may be required to ensure that the gas flow can be controlled in all situations. Temperature and level controllers are typically present in control situations that are not prone to fluctuate as widely or where the fluctuation can be readily and safely accommodated by the equipment. Therefore, such processes may be appropriate for control from a low bleed device, which is slower acting and less precise.

Safety concerns can limit the appropriateness of low bleed controllers in specific situations where any amount of bleeding is unacceptable. Emergency valves are often not controlled with bleeding controllers (e.g., neither low bleed nor high bleed), because it may not be acceptable to have any amount of bleeding in emergency situations (U.S. EPA, 2011a). Pneumatic controllers are designed for process control during normal operations and to keep the process in a normal operating state. If an Emergency Shut Down (ESD) or Pressure Relief Valve (PRV) actuation occurs,[15] the equipment in place for such an event is spring-loaded, or otherwise not pneumatically powered. During a safety issue or emergency, it is possible that the pneumatic

---

[15] ESD valves either close or open in an emergency depending on the fail safe configuration. PRVs always open in an emergency.

gas supply will be lost. For this reason, control valves are deliberately selected to either fail open or fail closed, depending on which option is the failsafe.

Costs

The costs described in this section are based on vendor research and information given in the appendices of the Natural Gas STAR Lessons Learned document on pneumatic controllers (U.S. EPA, 2006a). As Table 3-2 indicates, the average cost for a low bleed pneumatic is $2,553, while the average cost for a high bleed is $2,338.[16] Thus, the incremental cost of installing a low bleed device instead of a high bleed device is on the order of $165 per device. (Note: The ICF report assumed a cost of $3,000 to replace an existing high bleed pneumatic controller with a low bleed pneumatic controller based on industry comments (ICF, 2014).)

Table 3-2. Cost Projections for the Representative Pneumatic Controllers[a]

Device	Minimum cost ($)	Maximum cost ($)	Average cost ($)	Low Bleed Incremental Cost ($)
High bleed controller	366	7,000	2,388	$165
Low bleed controller	524	8,852	2,553	

[a] Major pneumatic controllers vendors were surveyed for costs, emission rates and any other pertinent information that would give an accurate picture of the present industry.

Monetary savings associated with additional gas captured to the sales line were estimated based on a natural gas value of $4.00 per Mcf (U.S. EIA, 2010).[17] The representative low bleed device is estimated to emit 6.65 tons, or 319 Mcf, (using the conversion factor of 0.0208 tons methane per 1 Mcf) of methane less than the average high bleed device per year. Assuming production quality gas is 82.8% methane by volume, this equals 385.5 Mcf natural gas recovered

---

[16] Costs are estimated in 2008 U.S. Dollars.

[17] The average market price for natural gas in 2010 was approximately $4.16 per Mcf. This is much less compared to the average price in 2008 of $7.96 per Mcf. Due to the volatility in the value, a conservative savings of $4.00 per Mcf estimate was projected for the analysis in order to not overstate savings.

per year (EC/R, 2011). Therefore, the value of recovered natural gas from one pneumatic controller in the production segment equates to approximately $1,500. While the owner of the transmission system is generally not the owner of the natural gas, the potentially lost gas still has value. The total value of the recovered gas from one pneumatic controller in the transmission segment is $1,375 assuming a natural gas value of $4.00 per Mscf and transmission natural gas is 92.8% methane by volume (EC/R, 2011).

3.1.3   Instrument Air Systems

Description

The major components of an instrument air conversion project include the compressor, power source, dehydrator, and volume tank. The following is a description of each component as described in the Natural Gas STAR document (U.S. EPA, 2006c), *Lessons Learned: Convert Gas Pneumatic Controls to Instrument Air:*

- Compressors used for instrument air delivery are available in various types and sizes, from centrifugal (rotary screw) compressors to reciprocating piston (positive displacement) types. The size of the compressor depends on the size of the facility, the number of control devices operated by the system and the typical bleed rates of these devices. The compressor is usually driven by an electric motor that turns on and off, depending on the pressure in the volume tank. For reliability, a full spare compressor is normally installed. A minimum amount of electrical service is required to power the compressors.

- A critical component of the instrument air control system is the power source required to operate the compressor. Because high-pressure natural gas is abundant and readily available, gas pneumatic systems can run uninterrupted on a 24-hour, 7-day per week schedule. The reliability of an instrument air system, however, depends on the reliability of the compressor and electric power supply. Most large natural gas plants have either an existing electric power supply or have their own power generation system. For smaller facilities and in remote locations, however, a reliable source of electric power can be difficult to ensure. In some instances, solar-powered battery-operated air compressors can be effective for remote locations, which reduce both methane emissions and energy consumption. Small natural gas powered fuel cells are also being developed.

- Dehydrators, or air dryers, are also an integral part of the instrument air compressor system. Water vapor present in atmospheric air condenses when the air is pressurized and cooled, and can cause a number of problems to these systems, including corrosion of the instrument parts and blockage of instrument air piping and controller orifices.

- The volume tank holds enough air to allow the pneumatic control system to have an uninterrupted supply of high-pressure air without having to run the air compressor continuously. The volume tank allows a large withdrawal of compressed air for a short time, such as for a motor starter, pneumatic pump, or pneumatic tools, without affecting the process control functions.

Compressed air may be substituted for natural gas in pneumatic systems without altering any of the parts of the pneumatic control. The use of instrument air eliminates natural gas emissions from natural gas powered pneumatic controllers. All other parts of a gas pneumatic system will operate the same way with instrument air as they do with natural gas. A diagram of a natural gas pneumatic controller is presented in Figure 3-1. A diagram of a compressed instrument air system is presented in Figure 3-2.

Applicability

The use of instrument air eliminates natural gas emissions from the natural gas-driven pneumatic controllers; however, these systems may only be used in locations with access to a sufficient and consistent supply of electrical power. Instrument air systems are also usually installed at facilities where there is a high concentration of pneumatic control valves and the presence of an operator that can ensure the system is properly functioning (U.S. EPA, 2006c).

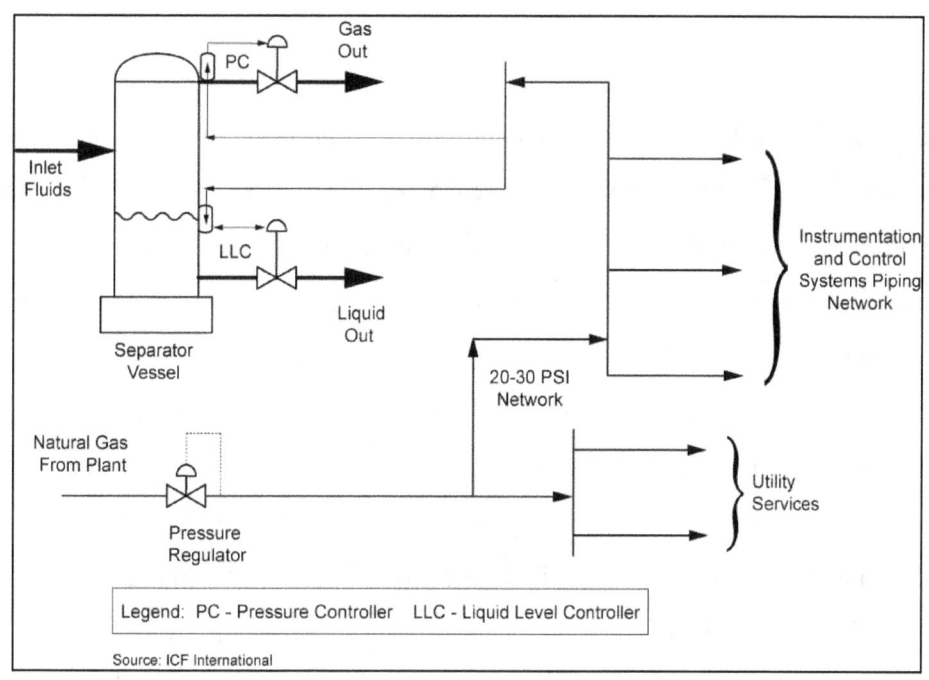

Figure 3-1 Natural Gas Pneumatic Control System

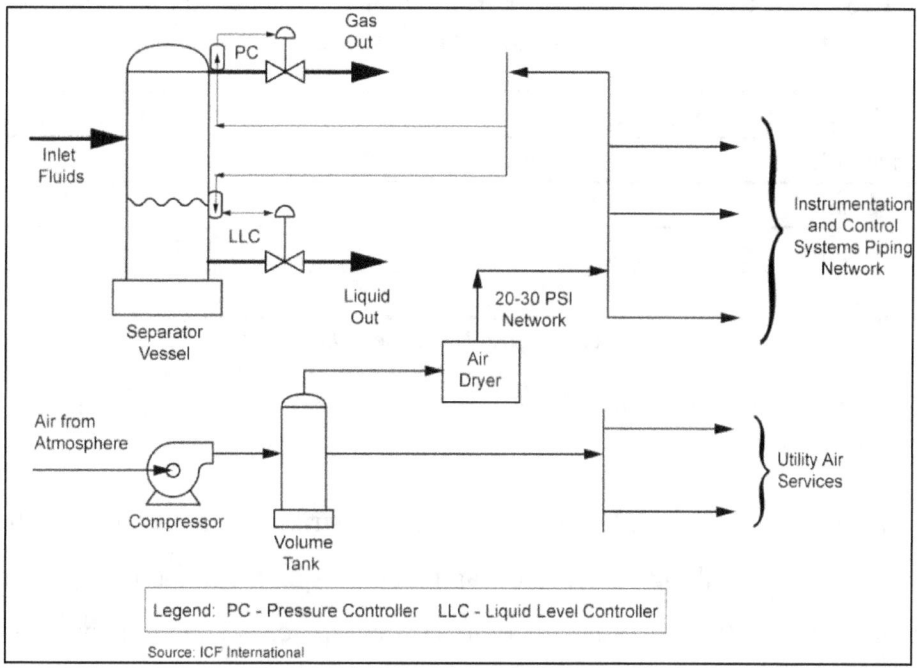

Figure 3-2. Compressed Instrument Air System

<u>Costs</u>

Instrument air conversion requires additional equipment to properly compress and control the pressured air. The size of the compressor will depend on the number of control loops present at a location. A control loop consists of one pneumatic controller and one control valve. The volume of compressed air supply for the pneumatic system is equivalent to the volume of gas used to run the existing instrumentation—adjusted for air losses during the drying process. The current volume of gas usage can be determined by direct metering if a meter is installed. Otherwise, an alternative rule of thumb for sizing instrument air systems is 1 cubic foot per minute (cfm) of instrument air for each control loop. As the system is powered by electric compressors, the system requires a constant source of electrical power or a backup pneumatic device. Table 3-3 outlines three different sized instrument air systems including the compressor power requirements, the flow rate provided from the compressor, and the associated number of control loops.

Table 3-3. Compressor Power Requirements and Costs for Various Sized Instrument Air Systems[a]

Compressor Power Requirements[b]			Flow Rate	Control Loops
Size of Unit	Hp	kW	(cfm)	Loops/Compressor
Small	10	13.3	30	15
Medium	30	40	125	63
Large	75	100	350	175

[a] Based on rules of thumb stated in the Natural Gas STAR document, *Lessons Learned: Convert Gas Pneumatic Controls to Instrument Air.*

[b] Power is based on the operation of two compressors operating in parallel (each assumed to be operating at full capacity 50% of the year).

The primary costs associated with conversion to instrument air systems are the initial capital expenditures for installing compressors and related equipment and the operating costs for electrical energy to power the compressor motor. This equipment includes a compressor, a power source, a dehydrator and a storage vessel. It is assumed that in either an instrument air solution or a natural gas pneumatic solution, gas supply piping, control instruments, and valve actuators of the gas pneumatic system are required. The total cost, including installation and labor, of three representative sizes of compressors based on assumptions found in the Natural Gas STAR document, "Lessons Learned: Convert Gas Pneumatic Controls to Instrument Air" are summarized in Table 3-4.

3.1.4   Mechanical and Solar-Powered Systems in Place of Bleed Controller

Description

Mechanical controls have been widely used in the natural gas and petroleum industry. They operate using a combination of levers, hand wheels, springs and flow channels with the most common mechanical control device being a liquid-level float to the drain valve position with mechanical linkages (U.S. EPA, 2006a). Another device that is increasing in use is electronic control instrumentation. Electricity or small electrical motors (including solar-powered) have been used to operate valves and therefore do not bleed natural gas into the atmosphere (U.S. EPA, 2006a). Solar control systems are driven by solar power cells that actuate mechanical devices using electric power. As such, solar cells require some type of backup power or storage to ensure reliability.

Applicability

Application of mechanical controls is limited because the control must be located in close proximity to the process measurement. Mechanical systems are also incapable of handling larger flow fluctuations (U.S. EPA, 2006c). Electric-powered valves are only reliable with a constant supply of electricity. These controllers can achieve 100% reduction in emissions where applicable.

Costs

Depending on supply of power, costs can range from below $1,000 to $10,000 for entire systems (U.S. EPA, 2006a).

Table 3-4 Estimated Capital and Annual Costs of Various Sized Representative Instrument Air Systems

Instrument Air System Size	Compressor	Tank	Air Dryer	Total Capital[a]	Annualized Capital[b]	Labor Cost	Total Annual Costs[c]	Annualized Cost of Instrument Air System
Small	$3,772	$754	$2,262	$16,972	$2,416	$1,334	$8,674	$11,090
Medium	$18,855	$2,262	$6,787	$73,531	$10,469	$4,333	$26,408	$36,877
Large	$33,183	$4,525	$15,083	$135,750	$19,328	$5,999	$61,187	$80,515

[a] Total Capital includes the cost for two compressors, tank, an air dryer and installation. Installation costs are assumed to be equal to 1.5 times the cost of capital. Equipment costs were derived from the Natural Gas Star Lessons Learned document and converted to 2008 dollars from 2006 dollars using the Chemical Engineering Cost Index.

[b] The annualized cost was estimated using a 7% interest rate and 10-year equipment life.

[c] Annual Costs include the cost of electrical power as listed in Table 3-3 and labor.

3.1.5 Maintenance of Natural Gas-Driven Pneumatic Controllers

Manufacturers of pneumatic controllers indicate that emissions in the field can be higher than the reported gas consumption due to operating conditions, age, and wear of the device (U.S. EPA, 2006a). Examples of circumstances or factors that can contribute to this increase include:

- Nozzle corrosion resulting in more flow through a larger opening.

- Broken or worn diaphragms, bellows, fittings, and nozzles.

- Corrosives in the gas leading to erosion or corrosion of control loop internals.

- Improper installation.

- Lack of maintenance (maintenance includes replacement of the filter used to remove debris from the supply gas and replacement of O-rings and/or seals).

- Lack of calibration of the controller or adjustment of the distance between the flapper and nozzle.

- Foreign material lodged in the pilot seat.

- Wear in the seal seat.

Maintenance of pneumatics can correct many of these problems and can be an effective method for reducing emissions. Cleaning and tuning, in addition to repairing leaking gaskets, tubing fittings, and seals, can save 5 to 10 scfh per device. Tuning to operate over a broader range of proportional band often reduces bleed rates by as much as 10 scfh. Eliminating unnecessary valve positioners can save up to 18 scfh per device (U.S. EPA, 2006a).

However, proper methods of maintaining a device are highly variable, thus, costs are variable based on labor, time, and fuel required to travel to many remote locations.

3.2 Available Pneumatic Pump Emissions Mitigation Techniques

There are several techniques that are currently being used to reduce emissions from pneumatic pumps. Table 3-5 provides a summary of these techniques for reducing emissions

50

from pneumatic pumps, which include chemical injection pumps and natural gas-assisted recirculation pumps.

3.2.1 Instrument Air Pump

Description

Circulation pumps in glycol dehydration processes and chemical injection pumps are often powered by pressurized natural gas at remote locations. As a result, these pumps vent natural gas to the atmosphere as part of their normal operation. To mitigate VOC and methane emissions, some companies are using instrument air to power these pumps. These companies have found that the use of instrument air increased operational efficiency, decreased maintenance and decreased costs, while eliminating emissions of methane and VOC (U.S. EPA, 2011b).

Applicability

Converting chemical injection pumps and glycol dehydration circulation pumps to instrument air can be applied to natural gas hydration operations across all gas industry sectors with excess capacity of its instrument air system. Because the systems are powered by electric compressors, they require a constant source of electrical power or a backup natural gas pneumatic device (U.S. EPA, 2011b).

Costs

The total cost to convert a natural gas pneumatic circulation pump to instrument air includes the installation of piping and an appropriate control system between the existing instrument air system and the glycol pump if the driver is independent of the circulation pump. If the driver is separated from the pump by O-rings, then the pump would need to also be replaced. The implementation capital costs are estimated to be $1,000 to $10,000, and the incremental operating costs are estimated to be $100 to $1,000 (U.S. EPA, 2011b). The potential annual

Table 3-5. Summary of Alternative Mitigation Techniques for Pneumatic Pumps

Option	Description	Applicability	Costs	Efficacy and Prevalence
Replace natural gas-assisted pump with instrument air pump (U.S. EPA, 2011b)	Circulation pumps in glycol dehydration units and chemical injection pumps are retrofitted with instrument air to drive the pumps (U.S. EPA, 2011b).	Facilities with excess capacity of the piping from the air compressor system to the pump or facilities that can install an air compressor system. Because the systems are powered by electric compressors, they require a constant source of electrical power or a backup natural gas pneumatic pump (U.S. EPA, 2011b).	The installation of instrument air or facilities that can install an air compressor system accounts for the bulk of the capital cost and typically ranges from $100 to $1,000 for chemical injection pumps (U.S. EPA, 2011b).	100% emission reduction, where applicable. The Natural Gas STAR reports typical annual methane savings to be 2,500 Mcf for glycol circulation pumps and 183 Mcf for chemical injection pumps (U.S. EPA, 2011b).
				The EPA does not have information on the prevalence of this technology in the field.
Replacement of natural gas-assisted pump with solar-charged direct current pump (U.S. EPA, 2011b)	In field settings, low volume natural gas pneumatic pumps can be replaced with solar-charged DC pumps (U.S. EPA, 2011b).	Low volume solar-charged pneumatic pumps are limited to approximately 5 gallons per day discharge at 1,000 psig. Large volume solar pumps are available with maximum output of 38 to 100 gallons per day at maximum injection pressures of 1,200 to 3,000 psig (U.S. EPA, 2011b).	The reporting partners for Natural Gas STAR stated a replacement cost of $2,000 per pump, including the solar panels, storage batteries and pump (U.S. EPA, 2011b).	100% emission reduction, where applicable. The Natural Gas STAR reports typical annual methane savings to be 182.5 Mcf per chemical injection pump conversion (U.S. EPA, 2011b).
				The EPA does not have information on the prevalence of this technology in the field.

52

Option	Description	Applicability	Costs	Efficacy and Prevalence
Replacement of natural gas-assisted pump with electric pump (ICF, 2014)	In settings where a constant supply of electricity is available, natural gas pneumatic pumps can be replaced with electric pumps (ICF, 2014).	These pumps require a constant source of electricity, thus, they are typically installed at processing plants or large dehydration facilities, which are normally equipped with electricity (U.S. EPA, 2011b).	Electrical pumps are estimated to cost roughly $10,000 per pump and the annual electrical usage cost was estimated to be $2,000 per year. (ICF, 2014)	100% emission reduction, where applicable.

The annual methane reduction from replacing pneumatic pumps with electrical pumps is estimated to be 5,000 Mcf (ICF, 2014).

The EPA does not have information on the prevalence of this technology in the field. |

natural gas savings are estimated to be 2,500 Mcf (U.S. EPA, 2011b) or $10,000 based on a natural gas value of $4.00 per Mcf (U.S. EIA, 2010). For chemical injection pumps, the implementation costs are the same, but the potential annual natural gas savings are estimated to be 183 Mcf per pump conversion (U.S. EPA, 2011b) or $732 based on a natural gas value of $4.00 per Mcf (U.S. EIA, 2010).

3.2.2   Solar Power Pump

Description

Solar power can be used to operate pumps located at remote sites where electricity is not available. These solar-powered pumps use electric power captured by solar panels to operate a DC-charged pump. Solar injection pumps can handle a range of throughputs and injection pressures. Low volume solar-charged DC pumps are limited to approximately 5 gallons per day discharge at 1,000 psig (U.S. EPA, 2011b). Large volume solar pumps are available with maximum output of 38 to 100 gallons per day at maximum injection pressures of 1,200 to 3,000 psig (U.S. EPA, 2011b). These pumps eliminate the methane and VOC emissions that would have resulted from the use of a pneumatic pump.

Applicability

These solar-powered pumps are generally used to replace low volume natural gas pneumatic pumps if sufficient sunlight is available to power the pumps and backup power is not required. These low volume pumps are typically used to inject methanol or corrosion inhibiters into producing wells and other field equipment. These chemical injection pumps are typically sized for 6 to 8 gallons of methanol injection per day. The large volume pumps can be used to replace gas-assisted circulation pumps for glycol dehydrators.

Costs

The Natural Gas STAR program reported the cost of replacing pneumatic pumps with solar-charged electric pumps to be approximately $2,000 per pump (U.S. EPA, 2011b). The solar panels and storage batteries are nearly maintenance free, and the solar panels have a life span of up to 15 years and the electric motors last approximately 5 years in continuous use (U.S. EPA, 2011b). The potential annual natural gas savings are estimated to be 2,500 Mcf or $10,000 based on a natural gas value of $4.00 per Mcf (U.S. EIA, 2010) for recirculation pumps (U.S. EPA, 2011b). For chemical injection pumps, the implementation costs are the same, but the potential annual natural gas savings are estimated to be 183 Mcf (U.S. EPA, 2011b) or $732 based on a natural gas value of $4.00 per Mcf (U.S. EIA, 2010). The ICF report estimates the cost of replacing chemical injection pneumatic pumps with solar-powered pumps to be $5,000 per pump with a natural gas savings of 180 Mcf per year (ICF, 2014) or $720 based on a natural gas value of $4.00 per Mcf (U.S. EIA, 2010).

3.2.3   Electric Power Pumps

Description

Electric power pumps are used to replace natural gas-assisted pneumatic used to recirculate glycol in gas dehydrators. These pumps eliminate the methane and VOC emissions that would have resulted from the use of a pneumatic pump.

Applicability

These pumps require a constant source of electricity, thus, they are typically installed at processing plants or large dehydration facilities, which are normally equipped with electricity.

Costs

Electrical pumps are estimated to cost roughly $10,000 per pump and the annual electrical usage cost was estimated to be $2,000 per year (ICF, 2014). The annual methane reduction from replacing pneumatic pumps with electrical pumps is estimated to be 5,000 Mcf (ICF, 2014) or $20,000 based on a natural gas value of $4.00 per Mcf (U.S. EIA, 2010).

4.0 SUMMARY

The EPA has used the data sources, analyses and studies discussed in this paper to form the Agency's understanding of emissions from pneumatic controllers and pumps and the emissions mitigation techniques. The following are characteristics the Agency believes are important to understanding these sources of VOC and methane emissions.

4.1 Pneumatic Controllers

- The majority of recent emissions estimates for pneumatic controllers have focused on methane emissions and not VOC emissions.

- The GHG Inventory data estimates that pneumatic controller emissions are 13% of overall methane emissions from the oil and natural gas sectors.

- Recent emission measurement studies have resulted in a wide range of methane emission factors for natural gas-driven pneumatic controllers. The studies all show that emissions can vary depending on sector (e.g., production, transmission, or storage) and the type of gas-driven pneumatic controller.

- Natural gas-driven pneumatic controllers are particularly useful in segments of the oil and natural gas industry that involve remote locations where electrical power is not available or reliable.

- Low bleed gas-driven controllers can replace high bleed gas-driven controllers in many, but not all, applications.

- Where a reliable source of electrical power is available, instrument air systems can replace natural gas-driven pneumatic controllers, and result in no methane or VOC emissions.

- Zero bleed, mechanical, and solar-powered controllers can replace continuous bleed controllers in certain applications, but are not broadly applicable to all segments of the oil and natural gas industry.

4.2   Pneumatic Pumps

- Pneumatic pumps in the oil and natural gas industry are used as chemical injection pumps and circulation pumps for glycol dehydrators. Pressure from the natural gas line is used to power these pumps and the natural gas is vented to the atmosphere.

- There are several mitigation techniques that can be used to reduce or eliminate emissions from pneumatic pumps and they include: instrument air pumps and electric pumps (both AC and DC powered).

- The 2014 GHG Inventory data estimates that pneumatic pump emissions are 16% of overall methane emissions from the natural gas production and processing sectors. The 2014 GHG Inventory estimated methane emissions from these sources to be 64,570 MT of methane for chemical injection pumps and 393,389 MT of methane for natural gas-assisted Kimray pumps. Chemical injection pumps at petroleum systems emitted 49,973 MT of methane, or around 3% of emissions from petroleum production.

- Natural gas-driven pneumatic pumps are particularly useful in segments of the oil and natural gas industry that involve remote locations where electrical power is not available or reliable.

- Where a reliable source of electrical power is available, instrument air systems are an effective replacement for natural gas-driven pneumatic pumps.

5.0   CHARGE QUESTIONS FOR REVIEWERS

1. Did this paper appropriately characterize the different studies and data sources that quantify emissions from pneumatic controllers and pneumatic pumps in the oil and gas sector?

2. Please discuss explanations for the wide range of emission rates that have been observed in direct measurement studies of pneumatic controller emissions (e.g., Allen et al., 2013 and Prasino 2013). Are these differences driven purely by the design of the monitored controllers or are there operational characteristics, such as supply pressure, that play a crucial role in determining emissions?

3. Did this paper capture the full range of technologies available to reduce emissions from pneumatic controllers and pneumatic pumps oil and gas facilities?

4. Please comment on the pros and cons of the different emission reduction technologies. Please discuss efficacy, cost and feasibility for both new and existing pneumatics.

5. Please comment on the prevalence of the different emission control technologies and the different types of pneumatics in the field. What particular activities require high bleed pneumatic controllers and how prevalent are they in the field?

6. What are the barriers to installing instrument air systems for converting natural gas-driven pneumatic pumps and pneumatic controllers to air-driven pumps and controllers?

7. Are there situations where it may be infeasible to use air driven pumps and controllers in place of natural gas-driven pumps and controllers even where it is feasible to install an instrument air system?

8. Did this paper correctly characterize the limitations of electric-powered pneumatic controllers and pneumatic pumps? Are these electric devices applicable to a broader range of the oil and gas sector than this paper suggests?

9. Are there ongoing or planned studies that will substantially improve the current understanding of VOC and methane emissions from pneumatic controllers and pneumatic pumps and available techniques for increased product recovery and emissions reductions?

6.0   REFERENCES

Allen, David, T., et al. 2013. *Measurements of methane emissions at natural gas production sites in the United States.* Proceedings of the National Academy of Sciences (PNAS) 500 Fifth Street, NW NAS 340 Washington, DC 20001 USA. October 29, 2013. 6 pgs. (http://www.pnas.org/content/early/2013/09/10/1304880110.full.pdf+html).

EC/R, Incorporated. 2011. Memorandum to Bruce Moore from Heather Brown. *Composition of Natural Gas for Use in the Oil and Natural Gas Sector Rulemaking.* June 29, 2011.

ICF Consulting. 1999. *Estimates of Methane Emissions from the U.S. Oil Industry.* Prepared for the U.S. Environmental Protection Agency. 1999.

ICF International. 2014. *Economic Analysis of Methane Emissions Reduction Opportunities in the U.S. Onshore Oil and Natural Gas Industries*. Prepared for the Environmental Defense Fund. March 2014.

Gas Research Institute (GRI)/U.S. Environmental Protection Agency. 1996a. *Research and Development, Methane Emissions from the Natural Gas Industry, Volume 5: Activity Factors*. June 1996. (EPA-600/R-96-080e).

Gas Research Institute (GRI)/U.S. Environmental Protection Agency. 1996b. *Research and Development, Methane Emissions from the Natural Gas Industry, Volume 6: Vented and Combustion Source Summary*. June 1996. (EPA-600/R-96-080f).

Gas Research Institute (GRI)/U.S. Environmental Protection Agency. 1996c. *Research and Development, Methane Emissions from the Natural Gas Industry, Volume 12: Pneumatic Devices*. June 1996. (EPA-600/R-96-080l).

Gas Research Institute (GRI)/U.S. Environmental Protection Agency. 1996d. *Research and Development, Methane Emissions from the Natural Gas Industry, Volume 13: Chemical Injection Pumps.* June 1996 (EPA-600/R-96-080m).

Gas Research Institute (GRI)/U.S. Environmental Protection Agency. 1996e. *Research and Development, Methane Emissions from the Natural Gas Industry, Volume 15: Gas-Assisted Glycol Pumps.* June 1996 (EPA-600/R-96-080o).

Roy, Anirban A. et al. 2014. *Air pollutant emissions from the development, production, and processing of Marcellus Shale natural gas*, Journal of the Air & Waste Management Association 64:1, pp 19-37. January 2014.

The Prasino Group. 2013. *Determining Bleed Rates for Pneumatic Devices in British Columbia; Final Report*. Prepared for the Science and Community Environmental Knowledge Fund. December 18, 2013. Available at http://www.env.gov.bc.ca/cas/mitigation/ggrcta/reporting-regulation/pdf/Prasino_Pneumatic_GHG_EF_Final_Report.pdf.

U.S. Energy Information Administration (U.S. EIA). 2010. *Annual U.S. Natural Gas Wellhead Price. Energy Information Administration. Natural Gas Navigator*. Retrieved online on 12 Dec 2010 at http://www.eia.doe.gov/dnav/ng/hist/n9190us3a.htm.

U.S. Energy Information Administration (U.S. EIA). 2012a. Total Energy Annual Energy Review. Table 6.4 Natural Gas Gross Withdrawls and Natural Gas Well Productivity, Selected Years, 1960-2011. (http://www.eia.gov/total energy/data/annual/pdf/sec6_11.pdf).

U.S. Energy Information Administration (U.S. EIA). 2012b. Total Energy Annual Energy Review. Table 5.2 Crude Oil Production and Crude Oil Well Productivity, Selected Years, 1954-2011. (http://www.eia.gov/total energy/data/annual/pdf/sec5_9.pdf).

U.S. Environmental Protection Agency (U.S. EPA). 2006a. *Options for Reducing Methane Emissions from Pneumatic Devices in the Natural Gas Industry*. Office of Air and Radiation: Natural Gas Star Program. Washington, DC, October 2006.

U.S. Environmental Protection Agency (U.S. EPA). 2006b. *Convert Pneumatics to Mechanical Controls*. Office of Air and Radiation: Natural Gas Star Program. Washington, DC, October 2006.

U.S. Environmental Protection Agency (U.S. EPA). 2006c. *Convert Pneumatic Controls to Instrument Air*. Office of Air and Radiation: Natural Gas Star Program. Washington, DC, 2006.

U.S. Environmental Protection Agency (U.S. EPA). 2011a. *Oil and Natural Gas Sector: Standards of Performance for Crude Oil and Natural Gas Production, Transmission, and Distribution Background; Technical Support Document for Proposed Standards*. Prepared for the Environmental Protection Agency. EPA-453/R-11-002. July 2011.

U.S. Environmental Protection Agency (U.S. EPA). 2011b. *Convert Natural Gas-Driven Chemical Pumps*. PRO Fact Sheet No. 202. Office of Air and Radiation: Natural Gas Star Program. Washington, DC, 2011.

U.S. Environmental Protection Agency (U.S. EPA). 2012b. *Technical Support Document: Federal Implementation Plan for Oil and Natural Gas Well Production Facilities. Fort Berthold Indian Reservation (Mandan, Hidatsa, and Arikara Nations), North Dakota. Attachment-FIP Emissions Control Cost Analysis from Operators*. 2012. EPA Region 8. EPA Docket No. EPA-R08-OAR-0479-0004.

U.S. Environmental Protection Agency (U.S. EPA). 2013. *Greenhouse Gas Reporting Program, Petroleum and Natural Gas Systems, Reporting Year 2012 Data.* Data reported by facilities as of September 1, 2013. (http://www.epa.gov/enviro/).

U.S. Environmental Protection Agency (U.S. EPA). 2014. *Inventory of Greenhouse Gas Emissions and Sinks: 1990-2012.* Climate Change Division, Washington, DC. April 2014. (http://www.epa.gov/climatechange/Downloads/ghgemissions/US-GHG-Inventory-2014-Chapter-3-Energy.pdf).